YOUR KNOWLEDGE HAS VALUE

AF145103

- We will publish your bachelor's and master's thesis, essays and papers

- Your own eBook and book - sold worldwide in all relevant shops

- Earn money with each sale

Upload your text at www.GRIN.com and publish for free

Reema Khetarpal-Kolge

Importance of Physical Activity and Nutrition- Prevention of Bone Density loss/Osteoporosis in Women post pregnancy

GRIN Publishing

Bibliographic information published by the German National Library:

The German National Library lists this publication in the National Bibliography; detailed bibliographic data are available on the Internet at http://dnb.dnb.de .

Imprint:

Copyright © 2014 GRIN Verlag GmbH
Print and binding: Books on Demand GmbH, Norderstedt Germany
ISBN: 978-3-656-88576-4

This book at GRIN:

http://www.grin.com/en/e-book/288246/importance-of-physical-activity-and-nutrition-prevention-of-bone-density

Importance of Physical Activity and Nutrition- Prevention of Bone Density loss/Osteoporosis in Women post pregnancy.

An Essay

Submitted to

PUBIC HEALTH FOUNDATION OF INDIA

In partial fulfillment of the requirements for
POSTGRADUATE DIPLOMA IN PUBLIC HEALTH NUTRITION
(via distance learning mode) 2014 BATCH

NAME OF THE CANDIDATE
REEMA KHETARPAL-KOLGE

Dated : 27 December 2014

Rationale

Pregnancy-associated osteoporosis was first reported more than forty years ago. At that time in 1955, there were reports of women who experienced vertebral fractures following delivery.

It appeared that if pregnancy-associated osteoporosis exists, it was probably quite rare. By 1996, eighty cases had been reported in the literature. It is difficult to adequately investigate, due to the inability to perform maternal radiologic exams. Pregnancy-associated osteoporosis tends to be identified in the postpartum period (56%) or the third trimester (41%).

Theoretically, pregnancy-associated osteoporosis is believed to occur because of the stress on maternal calcium stores and an increase in urinary calcium excretion. The body also responds to fetal calcium demands by increasing total 1,25-dihydroxyvitamin D levels. These two mechanisms are the reasons for the increased demand for calcium during pregnancy.

Physiologic changes during pregnancy include the third trimester estrogen surge and increased bone-loading due to weight gain. During pregnancy, the baby growing in its mother's womb needs plenty of calcium to develop its skeleton. (WHO studies)

This need is especially great during the last 3 months of pregnancy. If the mother doesn't get enough calcium, her baby will draw what it needs from the mother's bones but most women of childbearing years are not in the habit of getting enough calcium.

There is fact in the context when we look at studies that show pregnant women absorb calcium from food and supplements better than women who are not pregnant. This is especially true during the last half of pregnancy, when the baby is growing quickly and has the greatest need for calcium. During pregnancy, more estrogen, a hormone that protects bones is produced and therefore any bone mass lost during pregnancy is typically restored within several months after the baby's delivery.

However in recent times though there have been cases which indicate that women develop osteoporosis during pregnancy or breastfeeding, although it is rare. Transplacental calcium transfer occurs during pregnancy, especially during the last trimester. This is to meet the demands of the rapidly mineralizing fetal skeleton.

There is an obligate loss of calcium in the breast milk during lactation. Both these result in considerable stress on the bone mineral homeostasis in the mother. The maternal adaptive mechanisms to conserve calcium are different in pregnancy and lactation.

Studies have suggested that in spite of considerable changes in bone mineral metabolism during pregnancy, parity and lactation are not significantly associated with future risk for osteoporosis. However, the scenario is different in India.

The situation may not be the same as a significant proportion of pregnancies occur in the early twenties when peak bone mass is not yet achieved. Further, malnutrition, anemia and vitamin D deficiency are commonly encountered in this age group(NIN studies). This has an impact on future bone health of the mother. Loss of bone mass during lactation occurs mainly due to elevated PTHrP as well as hypo estrogenic state associated with high prolactin levels.

Current scenario in India

During puberty and adolescence, the skeleton takes up calcium avidly and builds up its reserves. This uptake of calcium into the bone is largely dependent on calcium and vitamin D nutrition, as well as exercise.

A good nutrition that is rich in calcium further helps to develop stronger bone density. It is highly advisable for girls at adolescent to have a good diet , proper exercise as this will help them to have reserves during pregnancy and for the growing needs of the fetus.

However , in India , the scenario varies. Traditionally in India, girls are not given the same social status as boys. Even though with education and knowledge, there are positive changes albeit it is not very significant but also limited to urban areas and more so in middle and high income social standings.

In low social income families and in rural parts of India , the tradition continues. Girls are not given healthy nutritious diets right from childhood. The girls are not encouraged to get proper exercise .more so in adolescence when the body is gearing for menstrual cycle.

Teen marriages are rampant which is followed by pregnancies and lactation . All these factors coupled with poor diet are responsible for bone loss during pregnancy as whatever reserve the mother has goes for the increasing demands of the fetus . The loss of calcium, poor absorption of nutrients and lack of minerals such as vitamin D all provide a major risk for osteoporosis post pregnancy.

An estimated 61% pregnant women are affected with osteoporosis in India.(NFHS 3) Studies have reported lower bone density among Indian women with osteoporotic fractures compared to their North American or European counterparts. There is a high prevalence of Vitamin d deficiency in Indian women right from their adolescence.(NIN studies)

Need for concern.

Post pregnancy osteoporosis causes fragile fracture mostly in vertebrae. The bone loss in lactating women is caused by calcium loss, decrease in estrogen level, and increase in PTHrP (parathyroid hormone related protein) level. Post pregnancy osteoporosis should be concerned, when we see a lactating woman with fragile fracture of the vertebrae.

Affects of osteoporosis are back pain, loss of height, and vertebral fractures, Hip pain and fracture of the femur are less common but not unusual.

Osteoporosis is a serious public health problem, this much studies have proven. But post pregnancy osteoporosis , even though rare, has to seen in the context of public health. If osteoporosis is not treated post pregnancy it can lead to serious health issues.

Osteoporosis is often called the "silent" disease, because bone loss occurs without symptomps. A common occurrence is compressionfractures of the spine. These can happen even after a seemin gly normal activity, such as bending or twisting to pick up a light object. The hunchback appeara nce of many elderly women, sometimes called "dowager's" hump or "widow's"hump, is due to th is effect of osteoporosis on the vertebrae.

Prevention

There is no cure for osteoporosis but with timely and alternate treatment one can reverse the effects .there are two major ways to prevent osteoporosis

- build strong bones with exercise
- a diet rich in calcium and vitamin D

Physical activity

Exercising regularly builds and strengthens bones Minimize bone   loss and possibly reduce the risk of broken bones. It Increases muscle strength Even if one already has osteoporosis, exercising can help maintain the bone mass. Exercise helps to maintain the amount and thickness of bones. Adequate physical activity early in life is important in reaching peak bone mass. Physical activities cause muscles and bones to work against gravity. Some examples of physical activities include

- Walking, Jogging, or running
- Stair climbing
- Jumping rope
- sports

Incorporating physical activity from early childhood ensures stronger bones throughout adult life. Young girls and adolescent girls should be encouraged to exercise right from their school days.

Nutrition

Proper nutrition is the one of the major prevention method for any disease. Early childhood nutrition will determine the health of an individual throughout his childhood, adolescence adult life. This is especially important for girls as in India, studies show that there is major lack of proper nutrition right from infancy. This leads to serious problems in adolescent age which is coupled with onset of menstrual cycle and prevalent teen marriages leading to early pregnancy. This leads to loss of bone density that may continue throughout adult life.

The deficiency of calcium and vitamin D is high among adolescent and adult pregnant women in India. Hence very important to have a proper diet in calcium and vitamin D rich foodsExperts recommend 1,500 milligrams (mg) of calcium per day for adolescents, pregnant or breastfeeding women.post pregnancy also the RDA for calcium is 1000mg a balanced diet should consist
of Milk, cheese, and yogurt. Other foods that are high in calcium are green leafyvegetables, tofu , shellfish, Brazil nuts, sardines, and almonds.

Supplementations Supplements vary in the amount of calcium them containthis helps in conditions where an individual may not acquire the minimum range of recommended calcium from the given diet.
Vitamin D helps the body absorb calcium. People can get vitamin D from sunshine with a quick (15-20 minute) walk each day or fromfoods such as liver, fish oil, and vitamin-
D fortified milk. 400 mg daily is usually the recommended amount.

Methods

Bone mineral density test

Bone density (**BMD**) is a clinical indirect indicator of osteoporosis and fracture risk. Bone density measurements are used to screen people for osteoporosis risk and to identify those who might benefit from measures to improve bone strength. Bone mineral density (BMD) between 1 SD and 2.5 SD lower indicates osteopenia. The assessment of BMD plays a vital role to identify the individual at risk for developing osteoporosis. Many of the published studies from various parts of India, including Ranu Patni's data, reported in this issue of the JMH, have shown lower values of BMD among young Indian women

- There has to be provisions made for population-based standards of peak BMD
- Policies to be formed for pregnant women to be assessed for bone mineral density .
- The results to be analyzed in terms of T-score for the incidences of normal bone mass, osteopenia, and osteoporosis.
- The same to be repeated post delivery and post lactation. This will help to gather information of all who are at risk of osteoporosis.
- Provisions and policies to implement BMD tests as mandatory in local health care centers, govt. hospitals, private clinics.

Nutrition

- Provision for diet rich in calcium and vitamin D during pregnancy as well as post pregnancy.
- Food fortification to be implemented keeping the needs of pregnancy and post pregnancy needs
- Cost effective fortified foods to be provided at local stores at all levels.

- Policies to be implemented to make a RDA recommended diet for all women during and after pregnancy.
- Training of local health workers to monitor the intake of the recommended RDA diet .
- Supplements for calcium and vitamin D to be provided at low cost at regular intervals during and after pregnancy.
- Training of health workers to monitor the intake of the supplements.
- Nutrition education to be made part of curriculum from an early age in all rural urban govt. and private schools.
- Policies to be made for education regarding nutrition to adolescent girls and counseling centers provided in schools local health centers.
- Focus of nutrition education to pregnant women , lactating women and women post delivery and lactation.

Physical activity

- Focus on physical activity from early childhood.
- Physical education to be mandatory to adolescent girls .
- Pregnant women to be trained for exercises like yoga , walking simple weight bearing exercise in health care centers.
- Local community group sessions by trained instructors at regular intervals during and after pregnancy.
- If for any reason facilities cannot be used at the local health centers, provisions made for home sessions at regular intervals.
- Walking camps to be held for communities and women to be encouraged to join them even after delivery.
- Post delivery counseling sessions to be provided at health care regarding all exercises , their nature and effects .
- Post delivery counseling sessions to provide information regarding onset of osteoporosis, its effects, it treatment and prevention

Expected Outcomes

Even though post pregnancy osteoporosis is rare , it nevertheless is an increasing concern for women and community as well. Low income and low socio communities can be benefited by the information, policies and provisions when this disease is addressed as a public health measure at all levels. Currently there are policies and provisions for food security , nutrition policies maternal policies but there is a dire need to focus specifically on onset of osteoporosis , its treatment , prevention. Osteoporosis is increasing gradually in women post pregnancy due to lack of proper nutrition , physical activity , improper food distribution, social norms, lack of education and knowledge regarding its effects and causes. Hence a policy implemented on the above methods will benefit adolescent girls, early pregnancies, post pregnancy care. Implementation of post pregnancy nutrition and physical exercise policies will help to impart knowledge at an early age regarding bone health, good nutrition, dietary supplementation, exercises to improve bone health. The policies would also provide a pathway for supplementation in early pregnancy as well as post pregnancy. The policies would implement health physical activity that would help to keep bone density mass strong and have high bone reserve thus minimizing the risk of developing osteoporosis.

References

New Delhi: ICMR: Published by Director General; 2010. Population based reference standards of Peak Bone Mineral Density of Indian males and females – an ICMR multi-center task force study; pp. 1–24.

Osteoporosis in Indians

N. Malhotra & A. Mithal

Osteoporosis during pregnancy and its management.

Smith R, Phillips AJ.

Scand J Rheumatol Suppl. 1998;107:66-7. Review.

Ranu Patni Normal BMD values for Indian females aged 20–80. J Mid-life Health.2010;2:70–73.

Bone remodeling and bone mineral density during pregnancy. Arch Gynecol Obstet

2003;268:309-16.

http://www.niams.nih.gov

http://www.ncbi.nlm.nih.gov

http://medical-dictionary.thefreedictionary.com/osteoporosis

http://www.nos.org.uk/

http://www.cdc.gov/

http://en.wikipedia.org/

www.osteoporosis.org.au

http://www.mamashealth.com/